Understanding a Millennium

by Gary Grimm and John Artman

Illustrated by Vanessa Filkins
Graphic Design by Seth Schaller
Cover by Vanessa Filkins

ISBN No. 1-56490-077-0

Printing No. 987654321

Understanding a Millennium is published jointly by **Gary Grimm & Associates** and **Pieces of Learning**. It is part of a grouping of products trademarked as ***Turning 2000™: Products for the New Millennium***. If you would like information on other products available, please write either company.

Gary Grimm & Associates
P.O. Box 378
Carthage, IL 62321-0378
1-800-442-1614
(217) 357-6763 FAX

Pieces of Learning
P.O. Box 340667
Dayton, OH 45434-0667
1-800-729-5137
(937) 427-3380 FAX
www.piecesoflearning.com

Understanding a Millennium

Table of Contents

What Is a Millennium?

We whose lives started before January 1, 2000, and continue past that date are slightly unique. We can say that we lived in two different millennia (plural) or millenniums (another accepted plural spelling). People born after 12:00 a.m. January 1, 2000, will live in just one millennium unless they live to be 1000 years old.

A millennium is a period of 1000 years. So, obviously, a new one doesn't occur very often. Even Halley's Comet or Hale Bop comes around more often than a millennium. Only those people born after 2875 will have a chance (by today's standards) to be like you and live a life that spans two millennia.

Actually, the new millennium should not occur until January 1, 2001. Why isn't it 2000? The reason is because our calendar of the western world dates back to the time of Julius Caesar. This is when BC and AD began. BC stands for before Christ. AD (anno Domini) means "in the year of the Lord." But a mistake was made. There was no year zero and there needed to be one. The years numbered 5BC, 4BC, 3BC, 2BC, 1BC, 1AD, 2AD, 3AD, etc.

See, they forgot to include a year zero. The first millennium started January 1, 1 AD, and so 1000 years would not pass until January 1, 1001. The second millennium should have started January 1, 1001, and it probably did.

SO WHY ARE WE CELEBRATING JANUARY 1, 2000?

The roll of the number from 1999 to 2000 is similar to the odometer on a car changing from 9999 miles to 10,000. There is something just a little bit exciting about it. In today's modern world we also have the media—television and news reporters. They have influence.

LET THE CELEBRATION BEGIN.

We Christian humans on planet Earth celebrate the new millennium New Year's Eve, December 31, 1999, and New Year's Day January 1, 2000. Did you notice the word *Christian* was used? That is important because the date, January 1, 2000, or January 1, 2001, is of little importance to followers of Judaism, Islam, and Buddhism. These religions use a lunar calendar instead of a solar calendar and do not follow the same dates or years as the Christian calendar. They will not be reaching a new millennium until our year 2001.

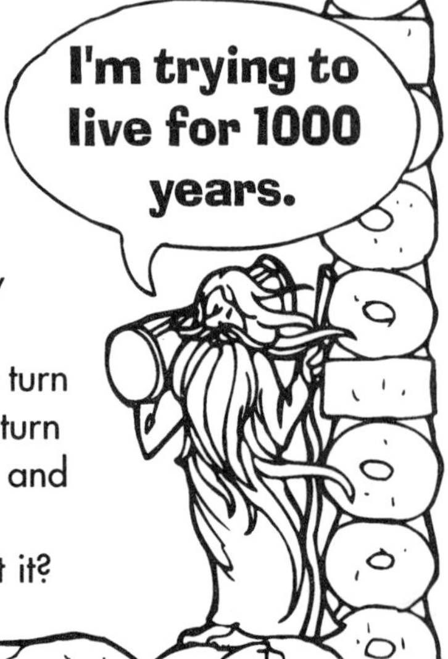

Seconds turn into minutes that turn into hours that turn into days. Days turn into weeks and months and a year. Years turn into decades and decades turn into centuries and every once in a great while the centuries accumulate and pass and we get to celebrate a new millennium.

It feels nice to be a part of this historic occasion, doesn't it?

HOW LONG IS 1000 YEARS?

Make marks (𝍸 III) to show each of the following:

What is your age?

What is your mother's/father's age?

How many years in a decade?

How many years in a century?

How old is your school building?

How old is America (born in 1776)?

How long has it been since Columbus discovered America?

How long is a millennium?

TIME FLIES WHEN YOU'RE HAVING FUN

bruncamels sheet sword hatt vaeh stinghome to do whit the sapings of emit:

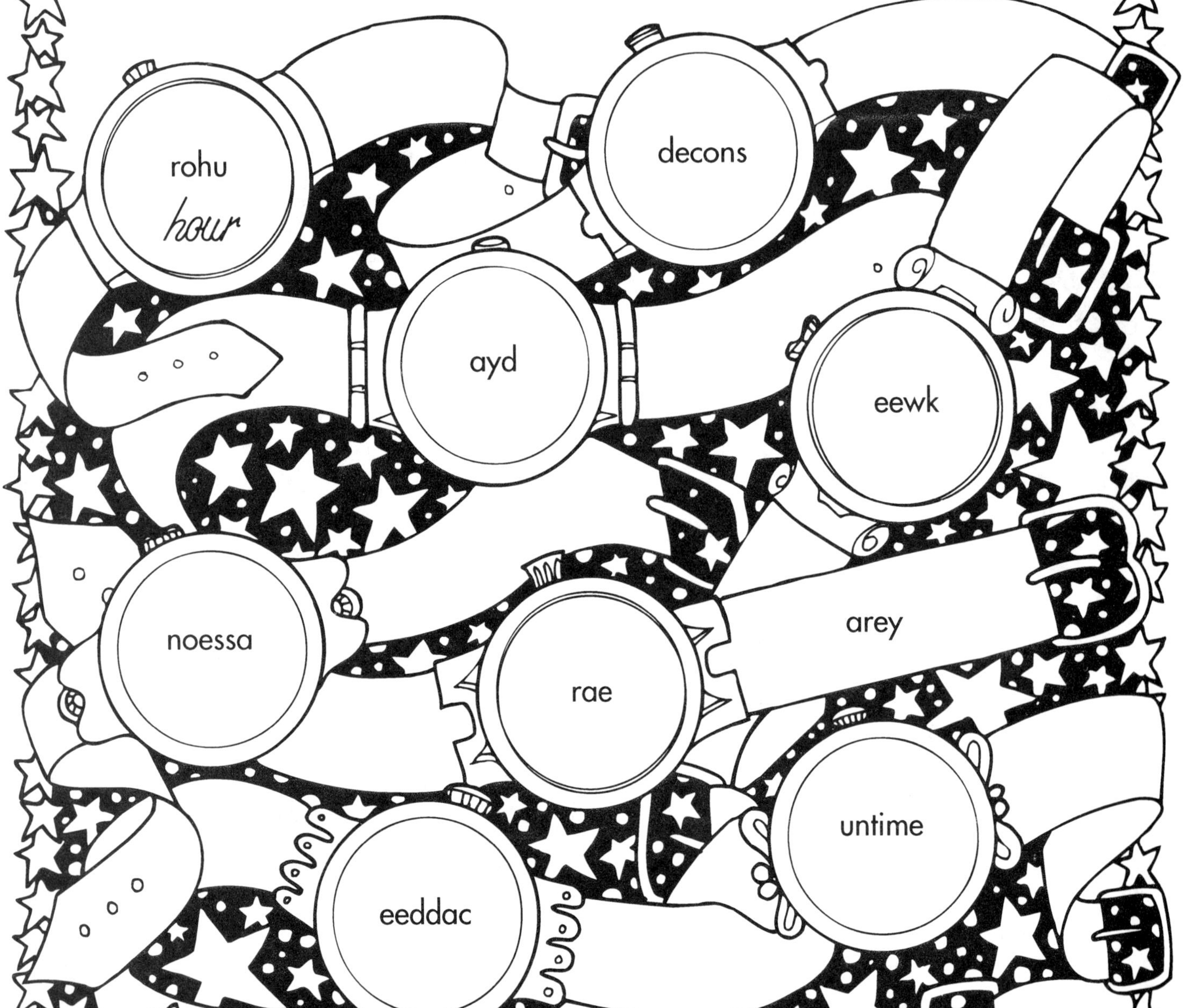

Can you define each of the words you unscrambled?

Hey, Just a Second!

Can you list ten things it takes just a second to do?

1. ______________________
2. ______________________
3. ______________________
4. ______________________
5. ______________________
6. ______________________
7. ______________________
8. ______________________
9. ______________________
10. wink
11. ______________________
12. ______________________
13. ______________________

How many

seconds in a minute? ______________

seconds in an hour? ______________

seconds in a day? ______________

seconds in a week? ______________

seconds in a month? ______________

seconds in a year? ______________

seconds in a decade? ______________

seconds in a century? ______________

seconds in a millennium? ______________

Work Space

When we think about how fast a second passes, it makes a millennium seem like a very long time. Actually a millennium is a short period of time when scientists tell us that the world is millions of years old. Think about it! How many millenniums in a million years? How many seconds in a million years? Just think about it. Don't even attempt to figure out the answer.

Can You Do It in a Minute?

There are 60 seconds in a minute. If it takes just a second to snap your fingers or wink, you should be able to do either 60 times in a minute. Try it. Can you snap your fingers exactly 60 times during a minute? Can you wink 60 times in a minute? Which can you do more of in a minute, wink or snap?

If there are 60 seconds in a minute, how many

minutes in an hour? ____________________ minutes in a year? ____________________

minutes in a day? ____________________ minutes in a decade? ____________________

minutes in a week? ____________________ minutes in a century? ____________________

minutes in a month? ____________________ minutes in a millennium? ____________________

What are some things you can do in a minute? LIST TEN

1. ____________________
2. ____________________
3. ____________________
4. ____________________
5. tie your shoes
6. ____________________
7. ____________________
8. ____________________
9. ____________________
10. ____________________
11. ____________________
12. ____________________
13. ____________________

Sometimes a minute seems much longer than it is. One of those times would be when the dentist is drilling. List some other times when one minute seems much longer.

____________________ ____________________

____________________ ____________________

____________________ ____________________

Before the Revolutionary War many American patriots were called minutemen. Why do you think these people got this name?

Hour by Hour

Do you think the hour segment is our most used measure of time? Many things happen by the hour. Most people earn an hourly wage. An appointment is scheduled at 9:00 or 9:30 (on the hour or half hour). We work so many hours a day. We say, "It is my lunch hour," even if it lasts less time. Television schedules are made on the hour and half hour. Can you think of a program that begins at 7:21 or 2:19? Of course not.

List some things it takes about an hour to do.

1. ______________________
2. ______________________
3. ______________________
4. ______________________
5. ______________________
6. ______________________
7. ______________________
8. ______________________
9. ______________________
10. ______________________
11. ______________________
12. ______________________
13. ______________________

If you could spend two hours doing exactly what you wanted to do, how would you pass the time?

____

____

____

____

____

How many

hours in a day? ______________________

hours in a week? ______________________

hours in a month? ______________________

hours in a year? ______________________

hours in a decade? ______________________

hours in a century? ______________________

hours in a millennium? ______________________

A Typical Day

1. A typical day is _______ hours long.
2. What are you usually doing when a typical day begins?

___

3. What are you doing when a typical day ends?

___

4. Complete the hourly schedule chart to the right. Write down what you are typically doing for each hour of a typical day. The date of the typical day is a Thursday in March.
5. What is the busiest hour of your typical day? ___________
6. What is your favorite hour of your typical day? ___________
   Why? ___
7. Is your daily schedule the same for every day of the week?
   Why? Why not? ___
8. Does your hourly schedule change from season to season? Is a typical Thursday in July the same as it is in March? ___
   Why? Why not? ___

___

9. How do you think a typical person spent a typical day in the year 1002?

___

10. Now think to the future. The year is 2921. What is a person's typical day like? How is it different from your mother's or father's typical day?

___

My Hourly Schedule on a Typical Day

12:00 - 1:00 ___
1:00 - 2:00 ___
2:00 - 3:00 ___
3:00 - 4:00 ___
4:00 - 5:00 ___
5:00 - 6:00 ___
6:00 - 7:00 ___
7:00 - 8:00 ___
8:00 - 9:00 ___
9:00 - 10:00 ___
10:00 - 11:00 ___
11:00 - 12:00 ___
12:00 - 1:00 ___
1:00 - 2:00 ___
2:00 - 3:00 ___
3:00 - 4:00 ___
4:00 - 5:00 ___
5:00 - 6:00 ___
6:00 - 7:00 ___
7:00 - 8:00 ___
8:00 - 9:00 ___
9:00 - 10:00 ___
10:00 - 11:00 ___
11:00 - 12:00 ___

What a Difference a Day Makes

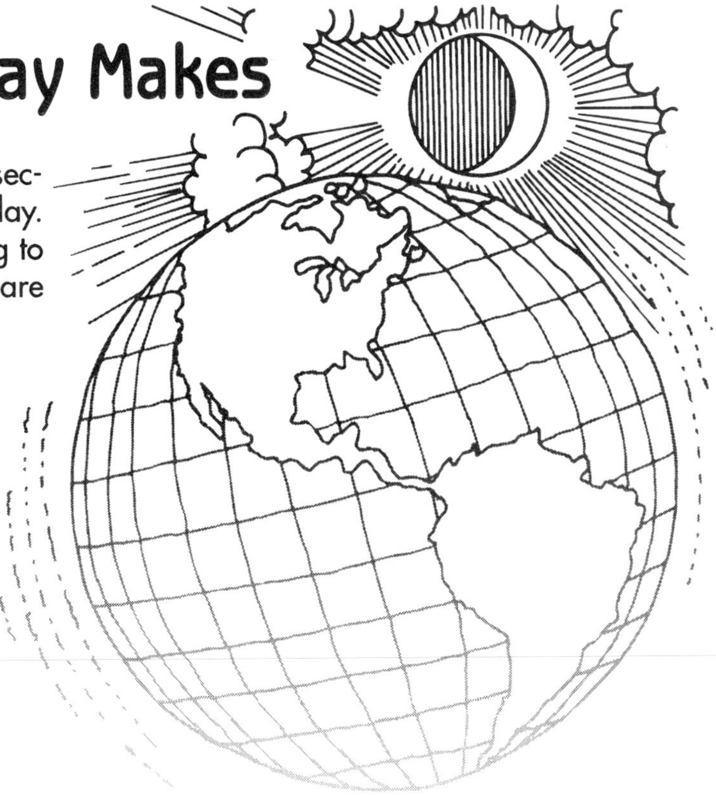

We talked about and completed work sheets on seconds, minutes, and hours. This one is about a day. However, when we think about a day we are going to be thinking about just the part of a day that we are awake.

What are some things it takes a day to do?

1. ______________________________
2. ______________________________
3. ______________________________
4. ______________________________
5. ______________________________
6. ______________________________
7. ______________________________

Seldom are there special seconds, minutes, or hours, but there are many special days. List five days that are special to you and tell why each is special.

Day/Date

______________ ____

______________ ____

______________ ____

______________ ____

______________ ____

What are five very important days that have occurred in the history of the world? Tell why each was very important.

Day/Date

______________ ____

______________ ____

______________ ____

______________ ____

______________ ____

Now, think about the future, your future. What do you think will be some very important days in your future? **Day/Date**

______________ ____

______________ ____

Have a Good Day

A good day is the same length as a typical day or a bad day. It just seems that a bad day lasts longer. Good, bad, or typical, a day is ______ hours long. How many minutes is that? ______ How many seconds is that? ______

	It's a Good Day When . . .	**It's a Bad Day When . . .**
weather		
parents		
brothers/ sisters		
school		
lunch		
free time		
friends	*invite me over*	
jobs/chores		
dinner		
TV		

See You Later, ALLIGATOR

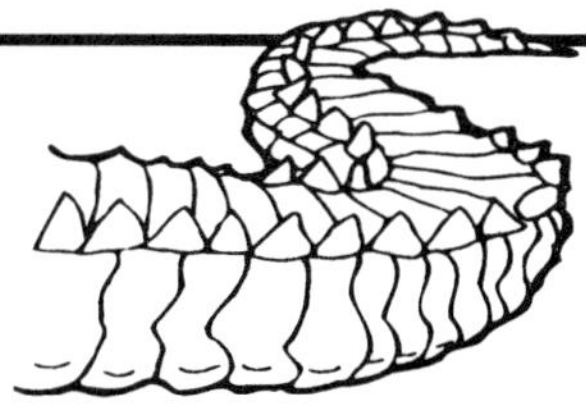

Clocks, calendars, watches, bells, and noontime sirens—our world is filled with devices that help us keep track of time. Time has been divided into seconds, minutes, hours, days. Everything should take place exactly, precisely on schedule. But it doesn't, especially when the human being is a factor. Human beings can be quite vague about time.

Below you will find some sentences. In each sentence you will find a reference to time that is less than precise. What does the underlined time reference mean to you? Follow the example.

1. See you later. ______________________________

   ____

2. Mother asked for help. "In a minute," I replied. *In a minute means in a few minutes— could be 5 minutes, 10 minutes, less than an hour*

3. I promised I would do my homework afterwhile. ______________________________

   ____

4. Our neighbor said she was going on a trip in a few days. ______________________________

   ____

5. Someday I would like to go to Chicago. ______________________________

   ____

6. I will take you to your friend's house pretty soon. ______________________________

   ____

7. Can you think of some more vague words/expressions that designate periods of time? ______________________________

   ____

8. Why do we humans use these vague references to time when there are so many precise ones available?

   ____

After you have determined what "in a little while" and some of the other vague terms mean to you, ask family members and friends to do the same. Compare what your version of "in a few minutes" is to those of your acquaintances.

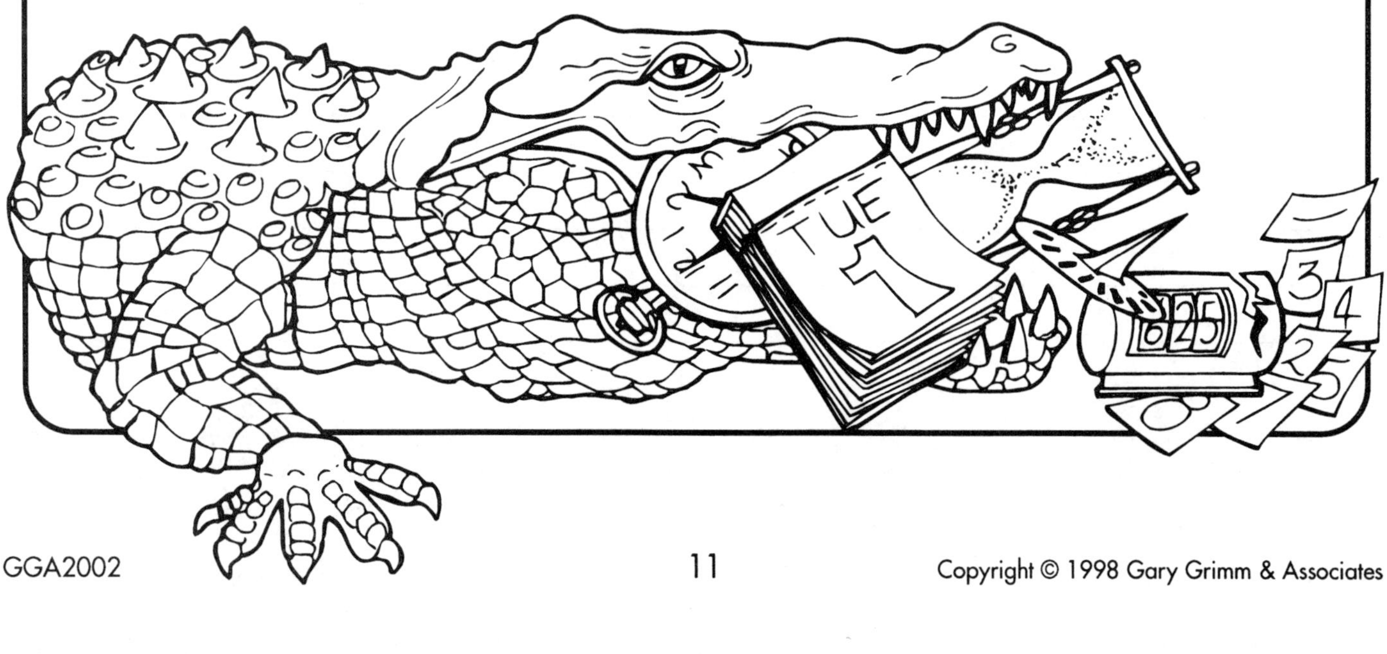

It Takes a Long, Long Time

Some things happen quickly: a clock ticks, you blink. Other things don't take very long to do: eat breakfast, walk to school, or get your teeth checked at the dentist.

It takes a little longer to do your homework (sometimes), watch a TV show, or play a ball game. However, it takes a long time for some things to happen. About how long does it take for

a seed to grow into a flower and bloom? ____________________

One season to change into another? ____________________

a year to pass? ____________________

a baby to become an adult? ____________________

a baby to become a grandparent? ____________________

a person to get a good education? ____________________

a new car to wear out? ____________________

a child to grow out of a pair of shoes? ____________________

a decade to pass? ____________________

a century to pass? ____________________

a generation of a family to occur? ____________________

However, none of the above takes a very long time when compared to how long it takes a millennium to pass.

List some things that take a long time to happen.

____

____

____

____

____

____

____

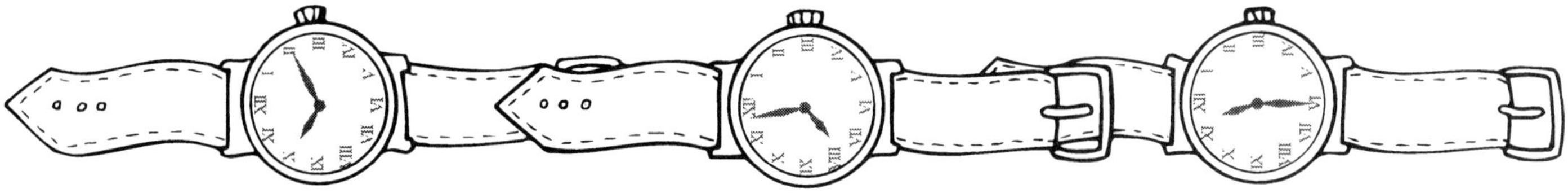

Number the following 1-10. One is the shortest amount of time it takes to happen and ten is the longest amount of time it takes to happen.

______flower to grow and bloom

______year to pass

______get an education

______outgrow shoes

______century to pass

______season to change

______baby becomes a grandparent

______new car to wear out

______decade to pass

______generation to occur

It is difficult to imagine just how long a millennium (1000 years) is. Do you think there is/are any living thing(s) on earth that have been around for at least 1000 years? List those things you can think of below.

Bring this list to class and be prepared to contribute to a class discussion. At the class discussion you and your classmates will share ideas. On the lines below write some of the ideas your classmates brought to the discussion session.

Days, Decades, Years, Centuries, Millenniums

The reason we have been thinking about seconds, minutes, hours, and days is because we are trying to discover just how long it takes for a millennium to pass. It's hard to understand this time period because no one has experienced one. Few people live a century. The average lifespan (in the United States) is 70-75 years. We know about weeks, months, and years because we can experience them. Maybe someday in a future millennium people will live to be 1000 years old. Do you think this will be possible?

So, we will continue manipulating numbers and trying to discover more about just how long a millennium is.

HOW MANY

days in a week? ____________________

days in a month? ____________________

days in a year? ____________________

days in a decade? ____________________

days in a century? ____________________

days in a millennium? ____________________

HOW MANY

weeks in a month? ____________________

weeks in a year? ____________________

weeks in a decade? ____________________

weeks in a century? ____________________

weeks in a millennium? ____________________

HOW MANY

months in a year? ____________________

months in a decade? ____________________

months in a century? ____________________

months in a millennium? ____________________

I'll bet you are glad these math sections are getting shorter and shorter.

HOW MANY

years in a decade? ____________________

years in a century? ____________________

years in a millennium? ____________________

HOW MANY

decades in a century? ____________________

decades in a millennium? ____________________

HOW MANY

centuries in a millennium? ____________________

Thinking About a Year

List some things most people do once a year.

take a vacation

What do you call something that happens / Example:

once a year?

twice a year?

four times a year?

every other year?

Give an example of each.

Relate each of the following numbers to a year.

365 | 52

4 | 6

once | a couple of

12

What worldwide event happens every four years?

What national event happens every four years?

List some things that happen once a year.

By the year 1600 most nations used January 1 as the first day of the new year. Prior to that most European nations used March 1 or March 25 as the first day of the new year. So the turn of the last millennium was either March 1 or March 25, 1000. Based on this information, when do you think the 3rd millennium will end and the 4th begin?

December November October September August July June May April March February January

The DECADES of a Person's Life

The average American male lives to be 72.3 years old and the average female lives to be 79 years old. So most people can expect to live at least seven decades—women almost a decade more. Complete the chart below based on the example used for the first decade.

	Age	Name	Most Important Events
First Decade	0 to 10	child	walking, talking, starting school
Second Decade	____ to ____	________	________
Third Decade	____ to ____	________	________
Fourth Decade	____ to ____	________	________
Fifth Decade	____ to ____	________	________
Sixth Decade	____ to ____	________	________
Seventh Decade	____ to ____	________	________
Eighth Decade	____ to ____	________	________

What is the most important decade in a human's life? ____________________

Why? ____________________

____

____

What is the most productive decade of an average human's life? ____________________

Why? ____________________

____

Glue a recent photo of yourself here.

____________________Name

__________Date ________Age

Draw yourself as you think you will look after at least two decades. You could make a light photocopy of your photo and age your youthful self.

____________________Name

__________Date ________Age

How do the phases of a human's life compare to those of other living things?
How many decades are in the life of an elephant or a turtle or a bird? What are these various phases called? What are some animals that do not live for a decade?

It Happened in the 20th Century

Listed below you will find twenty-one things that happened during the past 100 years. Which happened first? Which happened last? In the left column number the events 1-21 based on which happened first and which happened last. It might be fun to take the list home and have some family members help you decide.

After you have attempted to list the events in chronological order, go to work and research. In the first column of blanks to the right, write the year each event actually happened. In the third column number the chronological order of the events correctly.

______	you were born	______	______
______	Mickey Mouse born	______	______
______	Sojourner explores Mars	______	______
______	World War II ends	______	______
______	Man steps on moon	______	______
______	Vietnam War ends	______	______
______	first McDonald's opens	______	______
______	first color TV demonstrated	______	______
______	Elvis Presley dies	______	______
______	Clinton elected President	______	______
______	Wright Brothers' first flight	______	______
______	first Chevrolet car sold	______	______
______	World War I	______	______
______	first talking motion picture	______	______
______	Disney World opens	______	______
______	Alaska/Hawaii become 49th/50th states	______	______
______	first pizza in America	______	______
______	first Miss America crowned	______	______
______	JELL-O™ invented	______	______
______	Beatles come to America	______	______
______	regular airline service to Europe begins	______	______

A lot can happen in 100 years. Listed above are only twenty-one things. Some are of not much importance in comparison to the others. What do you think are the five most important things that have happened in the past 100 years?

____

____

Interviewing the Oldest Person I Know

We can learn from those older than we. One thing we can learn is what life was like when that person was young. Another thing we can learn is what has happened and what was new in the last 50+ years.

Have a chat with an elderly acquaintance. Include the following questions in your interview. Record the responses as the person talks, or use a tape recorder or video camera during the interview. The written form can then be completed at a later time.

The person I interviewed was

____

whose address is ______________________________________.

This person's date and place of birth is ______________________, ______________________.
date place

Tell me about the food(s) you ate when you were young.

a. What was a typical meal at your childhood home?

b. How was this meal prepared?

c. How are grocery stores different today?

d. Did you eat lunch at school? What did you have?

e. How often did your family eat out at a restaurant? What were the restaurants like?

f. When you were young, did they have

____ potato chips	____ catsup	____ videos
____ frozen pizza	____ candy bars	____ taco chips/salsa
____ a deli	____ cans of soda	____ packaged cheese

On other sheets of paper develop several more topics of questions for your interview.

Can you and the person you interview discover at least 25 things in his/her room or house that did not exist when he/she was a child?

HAPPY BIRTHDAY, 2000 A.D.

It is 11:59 p.m., December 31, 1999. In just one more second it will be a new millennium. Many things besides the millennium will change. On this page you will examine the many things that can change in just one second.

1. The millennium will change from the _________ millennium to the _________ millennium.
2. It will now be the _______ century instead of the _______ century.
3. It will be the _______ decade of the new century and millennium.
4. The year will change from _______ to _______.
5. The month will change from _______ to _______.
6. The day will change from ___________ to ___________.

   It will be the _______ week of the month, not the last.

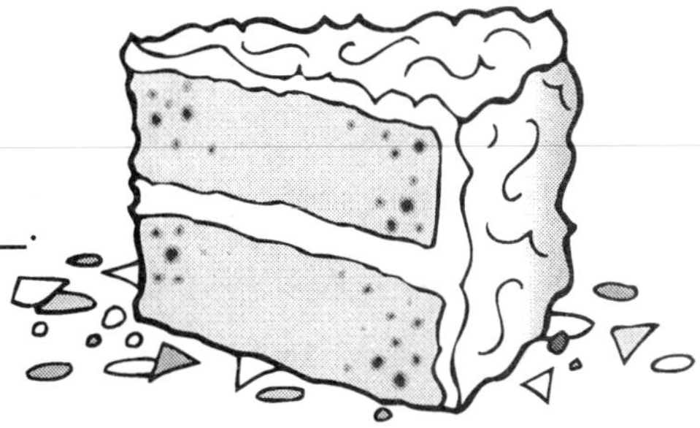

7. What are some other changes that will take place?

____

____

____

____

8. When the candles of a birthday cake are blown out, wishes are made. What wishes do you have for the new millennium, the next 1000 years? List some possibilities below.

____

____

____

____

____

____

____

____

____

Terminology & Time: A Review

You have completed several pages of exercises relating to time. You know that seconds turn into minutes, minutes turn into hours, hours turn into days, days turn into weeks, weeks turn into months, months turn into years, years turn into decades, decades turn into centuries, and centuries turn into millenniums. It should be easy for you to answer the following questions.

1. The year is 2001. What century is this?_______________.
2. It is 1980. How many decades until the next century?____________.
3. It is July 30, 1257. How many
   a. months until the next year?_______________
   b. months until the next decade?_______________
   c. years until the next century?_______________
4. The year is 1995.
   a. What century is it?_______________
   b. What decade is it?_______________
   c. What millennium is it?_______________
5. The year is 26 A.D.
   a. What century is it?_______________
   b. What decade is it?_______________
   c. How many years until the next decade?_______________
   d. What millennium is it?_______________
6. There are 3600 seconds in one _______________.
   There are 86,400 seconds in one_______________.
7. There are 10 years in a _______________,
   100 years in a _______________,
   and 1000 years in a _______________.
8. There are 12 months in a _______________,
   _______________ months in a decade and
   1200 months in a _______________.
9. How many weeks are there in a century? ___________
   How many weeks are there in a millennium? ___________
10. If we know how many weeks are in a millennium, we can determine the number of days in a millennium simply by
    ____.

Many digital clocks indicate if it is 12:10 a.m. or 12:10 p.m.
What do a.m. and p.m. stand for?

A Thousand Years

Now that's a long time.

WARNING! Don't look any place but right here and the next line.
How do you spell the word that means 1000 years? Write it on the next line.

____

OK! Now you can check your spelling by looking at another page in this book.

If you misspelled the word, write it twenty times on the reverse side of this paper. Then you can complete this sheet. If you spelled the word correctly, continue.

It takes many years for some things to happen.

Siberian tigers are an endangered species. There are less than ________ known to be living in the wild. Millions of dollars are being spent each year to try to keep these beautiful creatures from becoming extinct.

What do you think? Will Siberian tigers be saved or become extinct?

By the year 3000 how many Siberian tigers will still be alive in the world?____________

Some other things also happen slowly.

Wind and water erosion are constantly changing the shape and amount of coastline. What will happen to Florida? Already in some places there is hardly any beach left. Below in the left box draw what you think the state of Florida will look like in the year 3000.

In all the states bordering the Mississippi, wind, rain, and melting snow are causing erosion of the land. This dirt is ending up in the Mississippi River and flowing downstream to Louisiana. Each year the size of Louisiana is increasing. In the box to the right draw what you think the shape of the state of Louisiana will look like in the year 3000.

Florida 3000

Louisiana 3000

What are some other things that may take many, many years to happen? Prepare a list and bring it to a class discussion. Your teacher/leader will tell you when this discussion will take place. BE PREPARED!

Was it there?

1000 Years Ago

Did it exist?

Check yes or no. Was it there? Did it exist a millennium ago?

YES	NO		YES	NO	
______	_____	North America	______	_____	Egyptian pyramids
______	_____	United States of America	______	_____	bricks
______	_____	money (coins)	______	_____	the moon
______	_____	religion	______	_____	dinosaurs
______	_____	families	______	_____	books
______	_____	roads	______	_____	boats/ships
______	_____	English language	______	_____	Grand Canyon
______	_____	Mississippi River	______	_____	state of Mississippi
______	_____	dentists	______	_____	milk
______	_____	space	______	_____	war

List 10 more things that existed 1000 years ago.

____
____
____
____
____
____
____
____
____
____

List 10 things that are new (less than 25 years old).

____
____
____
____
____
____
____
____
____
____

It existed 1000 years ago
. . . but what was it like?

For each of the topics presented on this page, write a couple of comments. Your comments should focus on what you think. An example has been completed for you. One area is blank for you to select another topic and then make comments.

Music

Food

Shelter

Weapons/War

There were no motorized vehicles—only horses. Knights wore armor. Bows and arrows were the most common weapons.

Travel/Transportation

Work/Jobs

Money

Has Progress Been Made Since 1000?

What Progress Has Been Made?

Since the year 1000 there has been much progress. There have been discovery and invention. Many brilliant scientists and leaders have contributed to the betterment of civilization on earth.
Under each topic below list some of the important progress that has been made in the last millennium.

Medicine

Peace

Communications

Some inventions include telephone, fax, television, video recorder, and

Environment

Quality of Life

Personal Happiness

FADS

The dictionary says that a fad is a practice or interest followed for a period of time with exaggerated zeal. Zeal is enthusiasm.

In-line skating, bungee jumping, body piercing, and tattooing the body are some present day fads. How long will they last?

Fifty years ago most men wore their hair very short and few had beards or mustaches. During the 1980s and '90s long hair became popular. That was not new because long hair was popular 200 years ago. As we begin a new millennium, beards and mustaches are popular and short hair or a shaven head is once again the "in" thing. Fads come and go. Some remain and become part of our culture, and some last a short time and disappear never to be heard of again.

In 1891 James A. Naismith, a physical education instructor in Springfield, Massachusetts, was asked to create a team sport that could be played indoors during the winter. Naismith thought up thirteen rules for a game and asked a janitor of a gymnasium to nail two boxes to the gym's balcony to be used for goals. All the janitor could find were two half bushel baskets. Because of this Naismith called his invention *basketball*.

In December 1891, Naismith's class played the first game of basketball. There were eighteen students in Naismith's class so each team had nine players. They used a soccer ball and scored just one basket. Today, just over 100 years later, we have the NCAA Championships, the NBA, Chicago Bulls, Shaquil O'Neil, Michael Jordan, and the Utah Jazz.

Is basketball a fad?

How long before a fad becomes a part of the culture?

Can you list some fads of today that have a chance to last for a long, long time, maybe several hundred years?

How long do you think basketball will last? In the year 3000 will it be looked at as a fad of the 1900s?

With the help of your parents, neighbors, and grandparents can you name some fads that didn't last?

Can you and your classmates create a fad for this new millennium?

Baseball Thrives in the 3rd Millennium

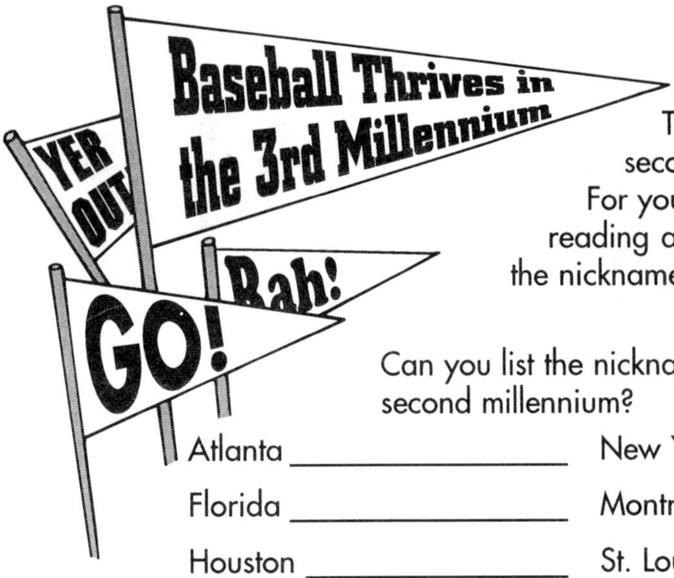

Everyone loves baseball. "Take me out to the ballpark," scream millions of people.

This popular pastime, invented in the latter part of the second millennium, is more popular than ever.

For you history lovers here is a quick quiz. Do you remember reading about baseball when it was in its infancy? Can you list the nicknames of the old, original National League teams?

Can you list the nicknames of these teams that existed at the end of the second millennium?

Atlanta ______	New York ______	Philadelphia ______
Florida ______	Montreal ______	Cincinnati ______
Houston ______	St. Louis ______	Chicago ______
Pittsburgh ______	Los Angeles ______	Colorado ______
San Diego ______	San Francisco ______	

And do you remember when the World League was formed in 2025? What were the nicknames of those teams?

Paris ______	London ______	Rome ______
Tokyo ______	Beijing ______	Mexico City ______
Rio de Jáneiro ______	Moscow ______	Sydney ______
Cairo ______	Kinshasa ______	Djakarta ______
Lima ______	Manila ______	Lagos ______

And now in 2525 we are about to have the first Universal League formed. Anyone in the universe can enter a contest to name the league members. If you are the lucky winner next year, you will be our guest at a home game in each of the new ballparks. After each team that will be a member, suggest what its nickname should be. If you are the winner, the universe is yours.

Mercury ______	Neptune ______	Jupiter ______
Mars ______	Pluto ______	Earth ______
Saturn ______	Uranus ______	

How do you think the game of baseball will change over the next 500 years?

____

____

____

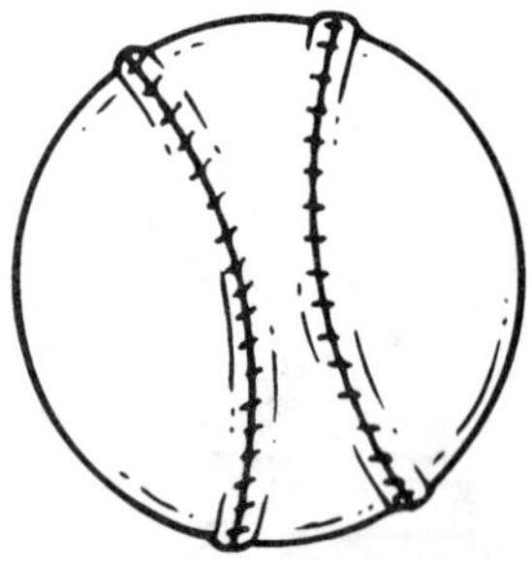

Will baseball exist 500 years from now? Baseball developed from the British game called *cricket*. Children in colonial times (1770s) played a game called *one-o-cat, two-o-cat*. That was only 225 years ago. The first professional baseball game was played in 1869.

Names Change

Below you will find a list of first names. During which millennium do you think each name was popular? Was it in the year 1, 1001, 2001, or is it a name that might be popular in 3001?

________ Ezekiel	________ Erik the Red	________ Trent
________ Moses	________ Salome	________ Warf
________ Bathsheba	________ Adolphus	________ Kyra
________ James	________ Troy	________ Moltin
________ Wade	________ Leia	________ Spartacus
________ Augustus	________ Catherine	________ Farrah
________ Darth	________ Lazarus	________ Claudius
________ William	________ Otto	________ Delilah
________ Catelin	________ Nefertitti	________ Lancelot

Which have changed more over the years, the names of men or the names of women? Why?

____

____

The names of places and objects also change. Iran was once Persia. The Congo, Zaire, and the Belgian Congo are all names of the same geographical region. Why do place names change?

____

____

A sofa, divan, davenport, and a couch are all basically the same thing—four words used to name one object.

In the space below list some other name changes.

____________ to ____________	____________ to ____________
____________ to ____________	____________ to ____________
____________ to ____________	____________ to ____________
____________ to ____________	____________ to ____________
____________ to ____________	____________ to ____________
____________ to ____________	____________ to ____________

LOOK Out for the Past; Here Comes the Future!

Things that happened in the past can affect us not only today but also in the future. What happened yesterday can affect tomorrow. What you do today can affect your future. What a nation, a group of people, does or doesn't do can have long lasting effects.

If you stay up late, it can cause you to have a bad day. If you don't learn to swim, you could jeopardize your future safety. If one nation invades and destroys another, it could be negative or positive or both.

What if . . .

1. you drop out of school before you graduate? ______________________________

   ____

2. you smoke a pack of cigarettes a day for thirty years? ______________________________

   ____

3. the Civil War had caused America to become two countries? ______________________________

   ____

   ____

4. you lie to your friends? ______________________________

   ____

What did you do yesterday that will have results today? ______________________________

____

____

____

What things can you do today that will make a better tomorrow for you? ______________

____

____

____

____

____

____

____

What has happened in the past that you think most affects who you are and what you can or can't do today?

List five things you have done in the past that should have a positive effect on your future.

1. ____________________
2. ____________________
3. ____________________
4. ____________________
5. ____________________

List five things you have done in the past that could have a negative effect on your future.

1. ____________________
2. ____________________
3. ____________________
4. ____________________
5. ____________________

List five things that are happening in the world today that should have a positive effect on its future.

1. ____________________
2. ____________________
3. ____________________
4. ____________________
5. ____________________

List five things that are happening in the world today that, if not corrected, could have a negative effect on its future.

1. ____________________
2. ____________________
3. ____________________
4. ____________________
5. ____________________

Is it possible that someone, somewhere, could do something today that would affect people who will be living in the year 2786? What could it be?

____

Photos for the Future

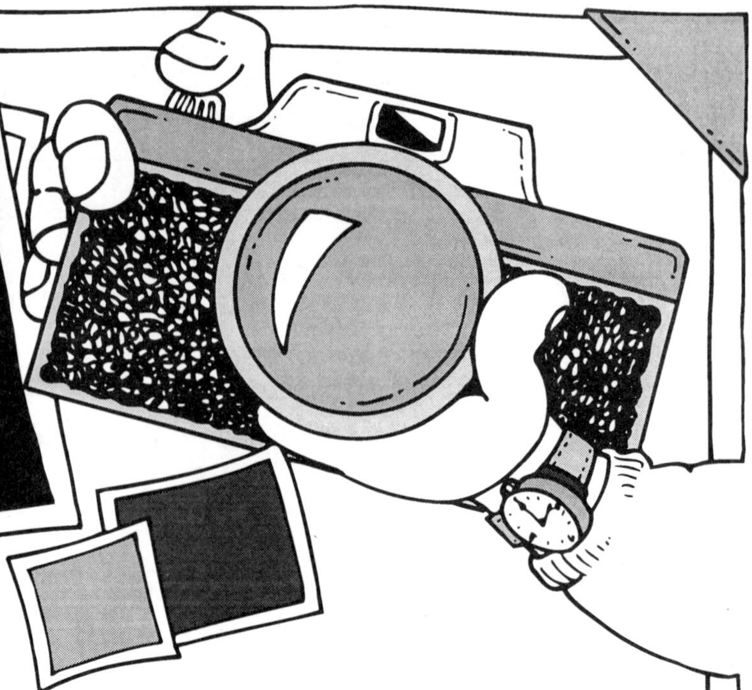

The National Historical Society has decided that they will help future historians. They will have someone in every community in the United States take photos of that community. These photos will be placed in airtight vaults and sealed. The vaults can be opened at the turn of the next millennium.

You have been chosen to take the photos of your community. You are limited to just ten photos. These ten photos should show/tell what your community was like at the beginning of the third millennium.

Remember they will not be viewed until the turn of the fourth millennium. List below what the ten photos will be.

1. ______________________________
2. ______________________________
3. ______________________________
4. ______________________________
5. ______________________________
6. ______________________________
7. ______________________________
8. ______________________________
9. ______________________________
10. ______________________________

Why did you choose the ten topics/subjects you did? ______________________________

____

____

____

If you could open a packet of photos from the past, what place in the world would you like to be able to see?

Or, if you opened a pack of ten photos of a North American community of the year 1000, what do you think you would see?

Learning About the Past

How do we learn about the past? Of course, we can read about the past in a book. But how did the author of the book you read learn about the past? If I wanted to know more about the 1960s, I would talk to my parents and grandparents and neighbors and people who were living during those years. And I would read books.

What do you think life was like at the turn of the last millennium? There is no one alive to ask. There is no one to talk to except a historian. Is that good enough? Does this person really know what the year 1000 was really like?

It is your job to learn as much about the beginning of the last millennium as possible. What was the world and life like 1000 to 1100? What could be your sources of information?

1. ____________________
2. ____________________
3. ____________________
4. ____________________
5. ____________________

Now it is your job to find out what life is like today in Burundi, Queensland, and Estonia. How will you gather this information.

1. ____________________
2. ____________________
3. ____________________
4. ____________________
5. ____________________

Which information was easier to gather? Why? ____________________

____

Which information is more accurate? Why? ____________________

____

Do you think learning about the past is easier than learning about the future? Is it possible to learn about the future?

Some Things Never Change

Yes, the weather changes from day to day and season to season if you stay in the same place. And, if you travel, the weather changes from place to place. But, if you lived in Omaha, Nebraska, for a year in 1005 and again in 2005, the weather would be basically the same.

List some other things that have changed very little during the last millennium.

1. ______________________________
2. ______________________________
3. ______________________________
4. ______________________________
5. ______________________________
6. ______________________________
7. ______________________________
8. ______________________________
9. ______________________________
10. ______________________________
11. ______________________________
12. ______________________________
13. ______________________________
14. ______________________________
15. ______________________________
16. ______________________________
17. ______________________________
18. ______________________________
19. ______________________________
20. ______________________________

What Progress Has Been Made?

The Olympic Marathon honors a Greek messenger who ran from the city of Marathon to Athens bringing news of victory in battle. After running approximately ______ miles and delivering the news, the messenger collapsed and died.

Today a phone call could be made or a fax sent. It would be faster and a life would be saved. The Greek runner would be amazed at the modes of communication that are available to us today.

What progress has been made? How do we communicate today? List some inventions that have improved communications during the past 1000 years.

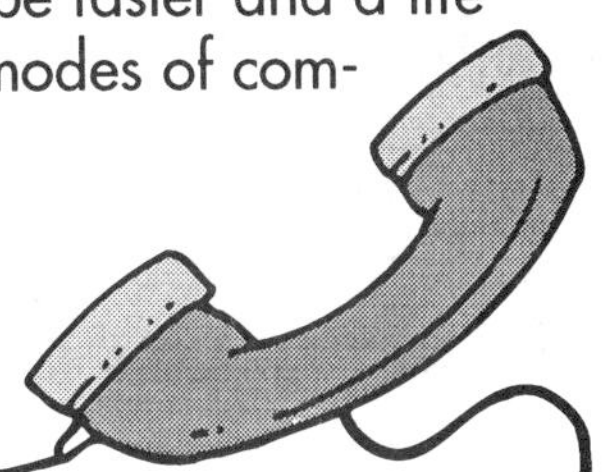

Inventions that have improved communications

The Invention	What the Invention Accomplished

Progress Report

Choose one of the following topics: EDUCATION, MEDICINE, TRANSPORTATION, PEACE. Write a report telling about some of the progress that has been made in this area in the last millennium.

Is all progress good? What has the growth of urban areas done to the environment? What other progress has had negative as well as positive results?

ADDRESSING THE FUTURE

Robin Hood lived in Sherwood Forest. That was his address. A Pilgrim farmer might have lived one mile south and a half mile east of Plymouth, Massachusetts (no zip code needed). Today, our address includes a street name and number or a post office box number, maybe an apartment number, and definitely a five-digit zip code. Soon a nine-digit zip code will be required. People who live in rural areas no longer live on Rural Route 1 or 2. They now live at an address like 1976 East County Road 2100.

Address the following two letters based on the following information.

Chauntelle Morgan lives in an underwater city ten miles west of Los Angeles. The year is 2447.

Prem Manoj Vivaldi lives on a space station tethered to the moon. The year is 2997.

Be sure to put your return address on each letter. Your address could change because of the year the letter is sent. Remember, progress continues.

Do you think the postal service as we know it will exist in 2500? Why? Why not? In 1516 the first successful postal system was established. It was between the cities of Berlin, Germany, and Vienna, Austria. On April 3, 1860, Pony Express service began in the United States. It was discontinued October 24, 1861, because the telegraph had been invented.

The Job Market Past, Present, Future

Jobs come and go. Fifty years ago no one worked at a Subway, Blockbuster Video, or at Disney World. These places did not exist then. Seventy-five years ago there were no air traffic controllers, flight attendants, or airports. Just think of all the new jobs related to the airline industry.

Will these jobs exist 100 years from now? You can be assured that in 2100 millions of people will work at jobs that do not exist today. What do you think these jobs will be?

You can also be assured that many jobs that exist today will not exist in the year 2100. List some jobs that existed in the year 1000, some that exist today, and some that you think might exist in the year 3000. Be sure to include some jobs that are now obsolete (for the year 1000) and some jobs that do not exist (for the year 3000).

1000	2000	3000
court jester	*magician*	
	computer graphics	
carriage driver	*pilot*	

Will schools and teachers be obsolete in the year 3000? If so, how will people learn?

Is There Life in Space? ___yes ___no

Don't check yes or no to express your opinion. Check yes or no to indicate if the question will be answered. Do the same for the questions. Check yes or no to indicate if we will have an answer to each of the following questions:

Will the polar ice caps melt?	___yes	___no
Will a cure for AIDS be found?	___yes	___no
Will there be peace on earth?	___yes	___no
Can we build cities underwater?	___yes	___no
Will a woman be elected President?	___yes	___no

In the space below list several additional unanswered questions that exist in today's world. Make a prediction about each. Ask four family members and/or friends what they think. Record each response. For example:

Someday will we be able to see the person as we talk to him/her on the telephone?

___yes ___no

Now give us your opinion. Yes or no, will the question be answered during the next millennium? If your response is yes predict what year the answer will be known.

Will the polar ice caps melt?	___yes	___no	___year
Will a cure for AIDS be found?	___yes	___no	___year
Will there be peace on earth?	___yes	___no	___year
Can we build cities underwater?	___yes	___no	___year
Will a woman be elected President?	___yes	___no	___year

Many times when one question is answered, it causes another to be asked, and so there are always unanswered questions. What do you think one future unanswered question might be?

EMPIRES COME AND GO!

An empire is a major political unit having a territory of great extent or a number of territories or peoples under a single sovereign authority. Greece, Rome, and Egypt all had great empires. Italy, Greece, and Egypt are no longer as important in the world as they once were. But their advanced civilizations contributed greatly to the development of the world as it exists today.

The United States of America is the most advanced culture in the world today. We are not an empire because we have elected leaders, and no one has complete power as the mighty Roman emperors once did. However, some people are concerned about the decline of the importance and influence of the United States in the world.

WHAT DO YOU THINK?

1. Will America be like the Greek, Egyptian, and Roman Empires and experience a decline in the quality of life and importance?

____

____

____

2. Do you feel there are already signs of decline in our society? If so, what are they?

____

____

____

3. If America declines during the next millennium, who do you think (what nation) will become the new leader of the planet Earth? ______________________
Why? ______________________________________

4. One unanswered question is: Is there life out there? Does another culture/civilization exist somewhere in space? Why? Why not? ______________________

____

____

5. America is over two hundred years old. How long will it survive? Will it last another hundred years? Will there still be a United States of America when the year 3000 arrives? If so, what might it be like? If not, why will it decline and disappear?

____

____

____

Continue on the reverse side.

The Latest in Fashion

Pictured on this page is some typical clothing worn by people around the year 1000. Styles surely do change. Why did these people dress like this?

There were no Wal–Marts™ or clothing stores 1000 years ago. How did people get their clothing?

On the back side of this sheet, draw some typical clothing that could be found in your closet. Think about color, fabric, and design. How are today's clothes different? Were there fashion designers in 1000? Were there choices? How much clothing did a person own?

The faces of the boy and girl below are typical for the year 3000. Complete the drawing showing how you think they might be dressed.

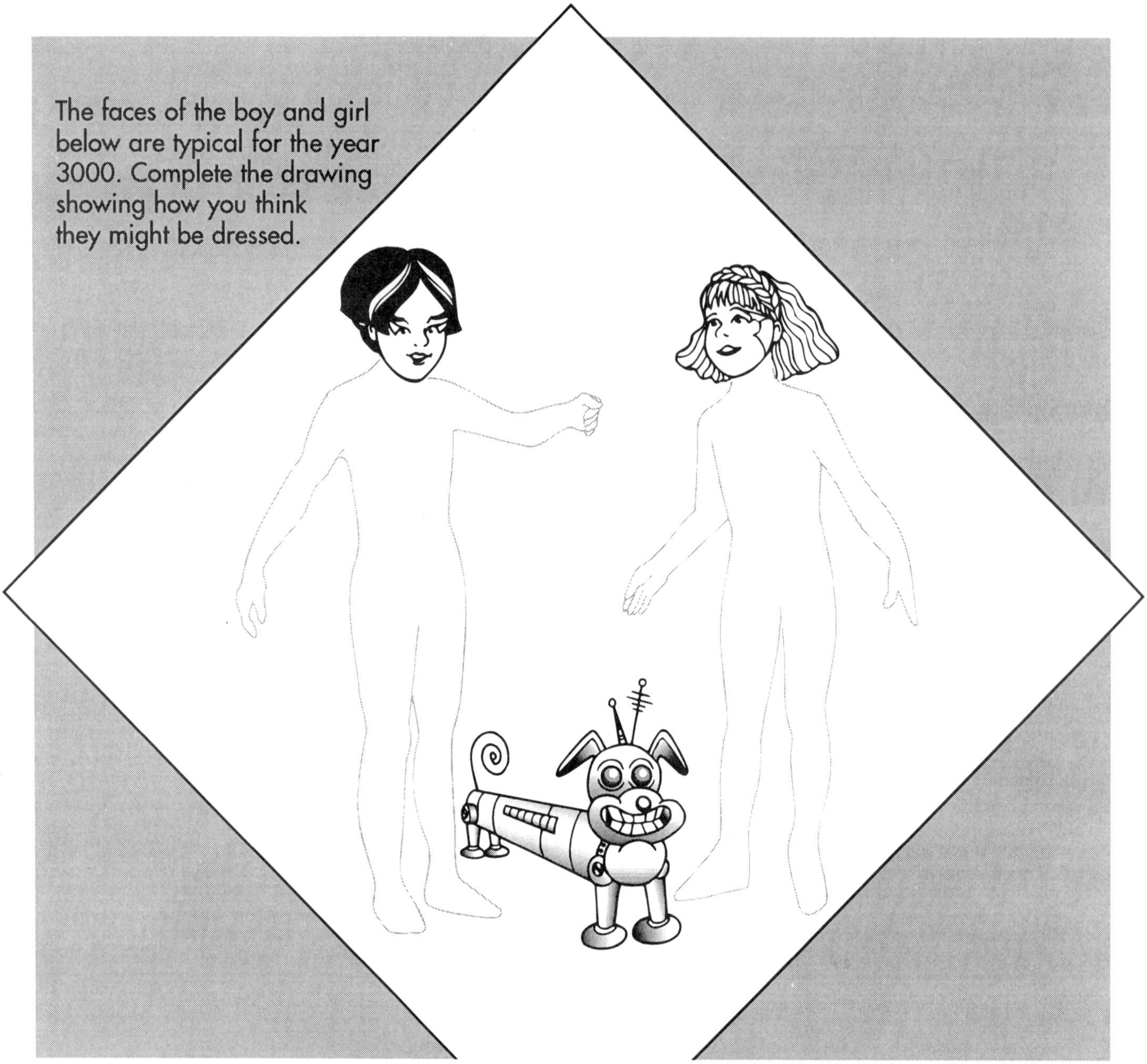

Fashion Forecast

On the lines below present some of your thoughts about acquiring, wearing, and caring for clothing as it might be in the year 3000.

The dictionary also defines *millennium* as "a period of great happiness or human perfection."

1. Has there ever been such a time in the history of the earth? What time in history was most like this definition of the word *millennium*?

2. What in today's world is perfect or almost perfect?

3. What is perfect (almost perfect) and imperfect about each of the following?

	PERFECT	IMPERFECT
you		
your community		
your country		
the world		

What are the components of happiness? Think, then list several. Bring your list to class. In class, one list that everyone can agree on will be created.

Transportation Time Line

Throughout time there have been many changes in methods of transportation. It is believed that it was around 6000 BC when man tamed animals like oxen and donkeys. 6000 BC was 8000 years or eight millenniums ago. Around 2700 BC the wheel was invented. Before then, walking was the main means of transportation.

The wheel was the first invention that led to the development of transportation. So much has happened since.

A Class Project

Your class will enjoy developing a transportation time line. It will serve not only as a quick reference to the development of the transportation industry but also can be used to illustrate that, as time passes, things develop at a faster and faster rate. Transportation, as we know it today, has been developed in the last one hundred years. Your completed time line will illustrate this.

What lies ahead? As you look at your completed time line and all that has happened in the past one hundred years, speculate what progress might be made in the new century and millennium.

Creating the Time line

Cut 8½" x 11" sheets of construction paper in half creating pieces that are 4¼" x 11". You will need twenty-three pieces this size. Number them in bold numbers 6000 BC, 2700 BC, and then 0, 100, 200, 300, etc., all the way to 1900 and 2000. See drawing below.

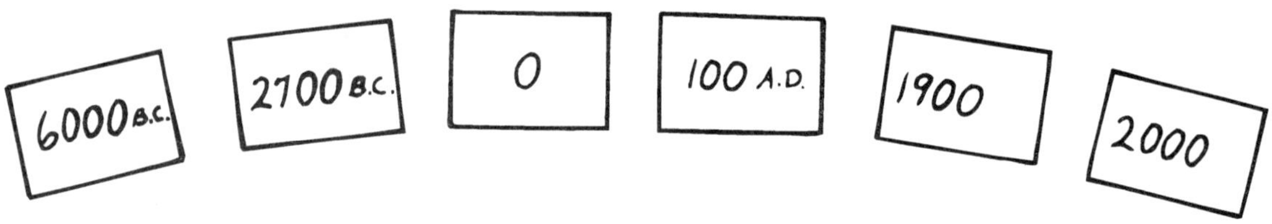

You will also need some wheel shapes. See pattern on the next page. One per student in your class and a few extras should be enough. On each of these a student will print a date and what, relating to transportation, occurred on that date. These will be placed on the time line in the appropriate places.

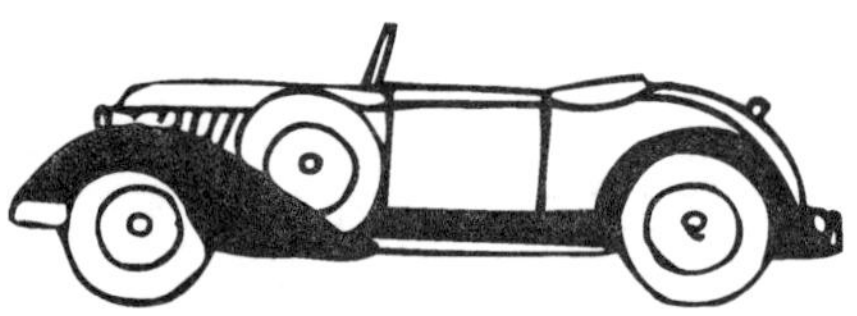

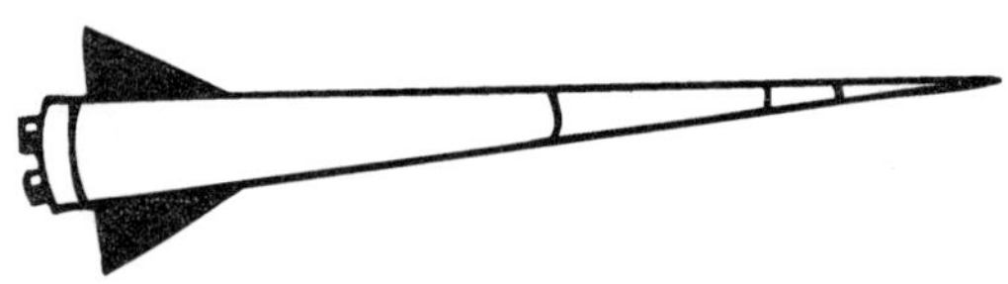

600 500 400 300 200 100 0 100 200 300 400 500 600 700 800 900 1000 1100 1200 1300 1400 1500 1600 1700 1800 1900 2000

Transportation Time Line *(cont.)*

The Topics

Assign the list of topics found below or allow students to choose. Each student should then do some research and determine when the topic became a part of the world's transportation system. Each student should prepare a short report to share with the class. The student should also complete a wheel to place on the time line.

automobile
hot air balloon
diesel locomotive
supersonic jet
water skis
bicycle
lunar rover
snowshoes
cable car
spaceship
covered wagon
subway
skis
pickup truck
space shuttle
snowmobile
aircraft carrier
parachute
escalator
elevator

airplane
pony express
steam locomotive
submarine
roller blades
sailboat
tractor
jeep
canoes/rowboats
steamboat
bus
taxicab
in-line skates
river barge
ultralight plane
horse
motorboat
van
golf cart

dirigible
sailing ship
jet airplane
dogsled
ambulance
atomic rockets
horse-drawn carriages
helicopter
gondola
semi truck
elevated train
motorcycle
wagon
jet ski
sled/toboggan
monorail
house trailer
ambulance
moving sidewalk

You and/or your students may wish to include some additional/alternative topics.

Students may cut pictures of methods of transportation from magazines and tape these to the time line to add color and interest.

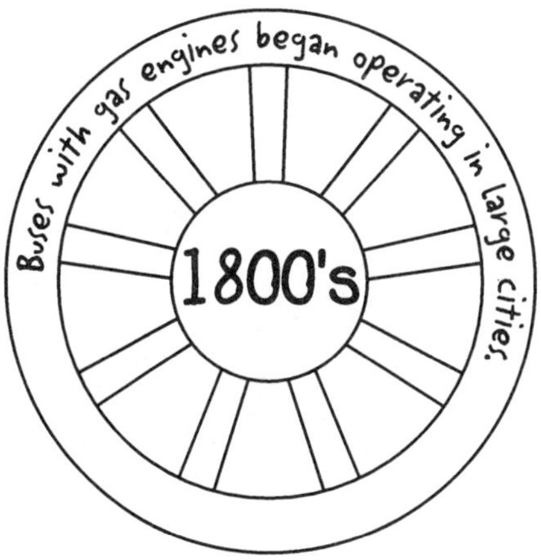

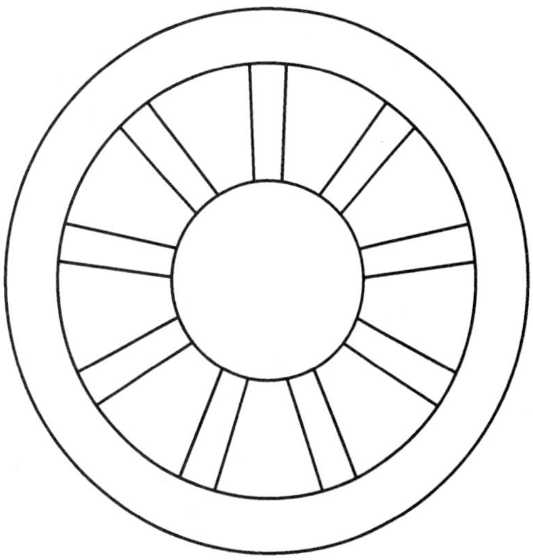

Television Time

How much television do you watch? Are you a frequent viewer or are you someone who watches only occasionally?

Log your viewing time for a one-week period using the chart below. Record the number of minutes you spend watching television before school, at school, and after school, each day for a week.

Record viewing time in the number of minutes watched. Calculate and record a total for each day and a total for the entire week. In your totals convert minutes to hours and minutes.

	Morning Wake-up to Noon	Afternoon Noon to 6:00	Evening 6:00 to Bed	Day's Total
Monday				
Tuesday				
Wednesday				
Thursday				
Friday				
Saturday				
Sunday				
			Week's Total	

How much time do you spend each week on each of the following?

sleeping ____________ daily ____________ weekly
classroom ____________ daily ____________ weekly
studying ____________ daily ____________ weekly
exercising ____________ daily ____________ weekly
playing ____________ daily ____________ weekly
working ____________ daily ____________ weekly
eating ____________ daily ____________ weekly

Before 1950 most people did not have televisions in their homes. What did they do instead of watching television?____

____

Review the findings of your survey of how you spend your time. How do you think this compares with someone your age who lived in the year 128 AD? How will someone your age spend his/her time in the year 2667?

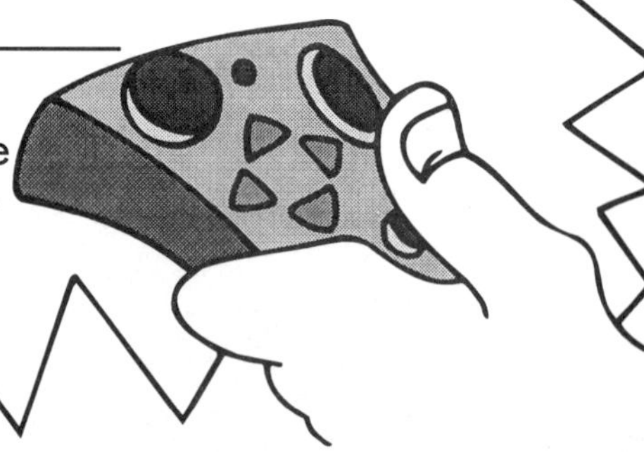

A Conception of Time

We do not know much about time. It is one of our greatest mysteries. We know we break time into segments, seconds, minutes, years, and millenniums. And in doing so we have measured time in different ways. We have measured time in the rising and setting of the sun, by the change of seasons, by the moon, hourglass, and clocks.

There is physical time, biological time, and geological time. All can be measured because they have length and mass. Biological time considers the rhythms of plants and animals. Geological time uses radioactive carbon to determine the age of the earth. But do any of these things really tell us what time is? Can we really grasp how long a millennium lasts?

One new theory concerning time is that the mass of the universe distorted time and pulled it in upon itself. So time is a sphere that has no beginning and no ending.

Here are some points to ponder with your students. Use these as motivation to the class or small group discussions.

1. What is time?
2. How do we know that time passes?
3. Aging shows that time is passing. What are the outward signs of aging that prove to us that time is passing?
4. What are some sayings that relate to time?

   Time waits for no man.

5. Many tales and stories begin with "once upon a time." What does that really mean?
6. What do the following terms have to do with time?

   International Date Line

   Prime Meridian

7. Why do we have various time zones in the United States and around the world?
8. Does time ever stand still?
9. If you are like most people there are times when time seems to drag and other times when time seems to speed by. Why is this? If time is dragging can you make it speed up?
10. If you were to place five items in a time capsule that would be opened in the year 3000, what would you place in it and why?

Answer Key

page 4:

hour	second
day	week
season	era
minute	year
decade	century
millennium	lifetime

page 5:
60 seconds in a minute
3600 seconds in an hour
86,400 seconds in a day
604,800 seconds in a week
2,419,200 seconds in a month
29,030,400 seconds in a year
290,304,000 seconds in a decade
2,903,040,000 seconds in a century
29,030,400,000 seconds in a millennium

page 6:
60 minutes in an hour
1440 minutes in a day
10,080 minutes in a week
43,200 minutes in a 30 day month
44,640 minutes in a 31 day month
525,840 minutes in a year
5,258,400 minutes in a decade
52,584,000 minutes in a century
525,840,000 minutes in a millennium

page 12:
About 3 months for seed to grow and bloom
90 days for a season to pass
365 days in a year
21 years for a baby to become an adult
40–50 years to become a grandparent
A minimum of 12 years for a good education
Answers will vary.
Less than a year to grow out of a pair of shoes
10 years for a decade to pass
100 years for a century to pass
20-25 years for a generation to pass

page 14:
7 days/week
30-31 days/month
365 days/year
3650 days/decade
36,500 days/century
365,000 days/millennium

4 weeks/month
52 weeks/year
520 weeks/decade
5200 weeks/century
52,000 weeks/millennium

12 months/year
120 months/decade
1200 months/century
12,000 months/millennium

10 years/decade
100 years/century
1000 years/millennium

10 decades/century
100 decades/millennium

1000 years/millennium

page 16:
Names for the various decades could be different from those listed.

age 0-10	child
age 11-20	teen/youth
age 21-30	young adult
age 31-40	adult/parent
age 41-50	adult/parent
age 51-60	middle aged/grandparent
age 61-70	retiree/grandparent
age 71-80	elderly/senior citizen

page 17:

Mickey Mouse	1928	Sojourner	1997
WWII ends	1945	Man on Moon	1969
Vietnam	1973	McDonald's	1948
color TV	1940	Elvis dies	1977
Clinton	1992	Wright Bros.	1903
Chevrolet	1913	WWI	1919
movie	1927	Disney World	1971
Alaska/Hawaii	1959	pizza	1905
Miss America	1921	JELL-O™	1902
Beatles	1964	airline	1939

page 19:
1. 2nd to 3rd
2. 21st instead of 20th
3. first
4. 1999 to 2000
5. December to January
6. 31st to 1st (first)

